MW00464318

Visualizing Nature

Visualizing Nature

—

Essays on Truth, Spirit, and Philosophy

Stuart Kestenbaum, editor

PRINCETON ARCHITECTURAL PRESS
NEW YORK

Contents

Ralph Waldo Emerson (1803–1882) published his essay *Nature* in 1836. It influenced thinkers and writers such as Henry David Thoreau and Margaret Fuller at the time and has continued to be an influence in American thought and writing. Many of Emerson's essays began as lectures. In 1833, in one of his first public lectures, he said, "Nature is a language and every new fact one learns is a new word; but it is not a language taken to pieces and dead in the dictionary, but the language put together into a most significant and universal sense. I wish to learn this language, not that I may know a new grammar, but that I may read the great book that is written in that tongue."[1]

The writers in *Visualizing Nature* continue to learn and speak this same language.

[1] Emerson, Ralph Waldo. 1959. *Early Lectures 1833–36*. Stephen Whicher, ed. Cambridge, Massachusetts: Harvard University Press.

Excerpts from Nature

Ralph Waldo Emerson

The wind sows the seed; the sun evaporates the sea; the wind blows the vapor to the field; the ice, on the other side of the planet, condenses rain on this; the rain feeds the plant; the plant feeds the animal; and thus the endless circulations of the divine charity nourish man.

—From Chapter 2, "Commodity"

But in other hours, Nature satisfies by its loveliness, and without any mixture of corporeal benefit. I see the spectacle of morning from the hill-top over against my house, from day-break to sun-rise, with emotions which an angel might share....

But this beauty of Nature which is seen and felt as beauty, is the least part. The shows of day, the dewy morning, the rainbow, mountains, orchards in blossom, stars, moonlight, shadows in still water, and the like, if too eagerly hunted, become shows merely, and mock us with their unreality....

The beauty that shimmers in the yellow afternoons of October, who ever could clutch it? Go forth to find it, and it is only a mirage as you look from the windows of diligence.

—From Chapter 3, "Beauty"

There is still another aspect under which the beauty of the world may be viewed, namely, as it becomes an object of the intellect. Beside the relation of things to virtue, they have a relation to thought. The intellect searches out the absolute order of things as they stand in the mind of God, and without the colors of affection....

The beauty of nature reforms itself in the mind, and not for barren contemplation, but for new creation.

—From Chapter 3, "Beauty"

[A true theory of nature and of man] always speaks of Spirit. It suggests the absolute. It is a perpetual effect. It is a great shadow pointing always to the sun behind us....

The happiest man is he who learns from nature the lesson of worship.

—From Chapter 7, "Spirit"

Introduction

Stuart Kestenbaum

SOME DAYS I TRAVEL BACK IN TIME. WHEN I GO FOR a walk up the hill from my house on the coast in Deer Isle, Maine, passing nineteenth-century buildings and ignoring the twentieth- and twenty-first-century utility poles and paved roads, I come to a field where the forest hasn't encroached. More than 150 years ago it was cleared for farming. From that opening I turn and look down on the village with its white Greek Revival capes, maple trees defining property lines, and the harbor that was once filled with sailing vessels transporting goods. The boats aren't there anymore—everything comes by truck or we drive somewhere to buy it—but the tide is still as dependable, and the cumulus clouds parade across a blue New England sky, just as they did in Ralph Waldo Emerson's day.

On these walks I often remember that Emerson described walking past "twenty or thirty farms" and "the woodland beyond. But none of them owns the landscape. There is a property in the horizon which no man has but he whose eye can integrate all the parts... This is the best part of these men's farms, yet to this their warranty-deeds give no title."

Slipping back or moving forward in time, I imagine that the two of us could be walking side by side, admiring this landscape, although I would have a lot of explaining to do when we came across stone walls that once marked farm fields now running through second-growth woods without a cow or sheep in sight. And that would just be the start of it. From there I'd have to move through nearly two centuries of technology and history with its achievements and horrors woven together.

Whatever historical differences we'd encounter on our walk, we would still meditate on the horizon, searching for ways to integrate all the parts. Emerson could apprehend the cosmic unity in nature that could make him a "part or a particle of God." He, and the men and women of his time, stood closer to the center of their universe than we can imagine ourselves today. We have more science, but we are less certain about our place in the cosmos and increasingly aware of our planet's fragility, both in the natural world and in our own societies.

Nature still speaks to us—how could it not—but we inhabit a more battered world and witness with alarm the breakdown of the elemental connections that inspired awe in Emerson. There are days when we wonder how we can carry on.

Yesterday, the last day in May, I sat in the sun in my backyard in the year that a global pandemic has, for the time being, turned our world upside down. The news is full of opinion pieces and articles asking, "Is this the end of—fill in the blank—as we know it?" The wind picked up and suddenly the air was full of dandelion seeds, floating and swirling like snow flurries, each looking for a home in the earth. A winged maple seed fell from a tree, landing on the page of the book I was reading. The world was going about the work at hand. Survival and renewal were literally in the air.

In *Visualizing Nature* we invited writers from varied backgrounds and disciplines to reflect on the themes in Emerson's essay "Nature" and tell us what they encounter in today's world. How does the natural world speak to us and how do we listen? Here is the news from second-growth forests, from deserts and marshes, from coral reefs, from trees that live for hundreds of years, from the waves breaking at the Jersey Shore. Here is the news that harmony may still be here.

Essays

On Discipline

Alison Hawthorne Deming

"NATURE IS A DISCIPLINE," EMERSON WRITES.

What could he mean by this? Discipline for me means order and routine, practice and repetition. I need discipline to keep up my exercise routine, especially these days when the Sonoran Desert soars over 100 degrees at the end of April. I need discipline to practice piano when I want to learn to play "What a Wonderful World" with spicy jazz chords. I need discipline to stay at the desk when ideas skitter away like mercury from a broken thermometer. I need discipline to stay home and wash my parched hands yet again while COVID-19 speeds through my city.

I do not, however, need to be disciplined for falling short. Nor do I need to take vows of poverty and celibacy for a life of religious discipline. And, no, I do not need to try the sexual high jinks of bondage and discipline. Nature is inspiration, nature is home, nature is commodity for body and soul, nature is wound and healing, decay and fruition. But discipline? Emerson, well educated in the classics, reaches back to the Latin sense of the word: "the instruction of disciples." When I go there with him, feeling myself instructed by nature, I fall into sympathy with his thought. I am a disciple of nature, and this idea is at the core of my environmental philosophy.

Emerson is tricky to read. He can describe thought with such detail and veer into such vertiginous tangents that reading him can feel like playing a video game of the mind. What's the target? Who knows. But it makes for a fabulous ride, if you can hold on. Emerson jolts a twenty-first century reader out of the materialistic mindset of her time back to the center of thought and writing that was Concord, Massachusetts, in the nineteenth century.

Think Emerson, Thoreau, Hawthorne, Fuller, Alcott. And the good citizens who turned out in crowds to hear their lectures. Here, the American land gave instruction to its inhabitants that nature was to be of use not only in the kitchen and barnyard but also, and most importantly, in the world of thought, in the life of the spirit. The discipline of nature is to transfer experience into thought "by perceiving the analogy that marries Matter and Mind." It is a short step from these words of Emerson's to Muriel Rukeyser's "Breathe in experience, breathe out poetry."

When Emerson visualizes nature, he sees the invisible world of "Universal Being." Our century demands that we visualize nature grounded in data. Sure, we feel a nudge of spirit in the exuberance of spring bloom, expanse of silent sea, lift of aged mountains. But our knowledge of the state of the planet calls us back to data: parts per million of carbon in the atmosphere, number of species lost and losing, cultures eroded or awash in higher tides, oceans becoming fallow from rising temperature and acidity. This is the new knowledge of our century.

Nature is a victim of human excess.

We read the graphs and numbers. We lament that political and economic interests outcompete science and moral intelligence in our rising to the challenge. It's no surprise that data alone does not change human behavior. Data is both the strength and weakness of our perception of nature. Emotion drives human behavior, and it is very hard to have a feeling about a graph or chart. Threats to the planet's well-being do not necessarily feel like threats to one's personal well-being. The COVID-19 pandemic offers a lesson in learning to expand the sphere of our moral regard, as we see that most people are willing and able to make sacrifices for the sake of the common good if they feel vulnerable.

I gaze up from my desk at the mesquite tree newly leafed blowing in the desert breeze. Yesterday I saw the great horned owl land on one of its branches and perch there for hours contemplating the day. Its head turned this way and that. Occasionally it turned to look at me, then turned away, keeping its back to me, content in its thoughts. The tree and the owl and the day and my gaze all seemed one thing held together by a sliver of time and place.

Today, the wind has picked up, making the mesquite's hair blow like it's on a wild convertible ride, its lithe limbs made for this swimming in air. What would it be to sway with the wind while firmly rooted, to feel after winter's dormancy the stipule emerge and then the leaf and then the flow of sap and the start up of the machinery that turns sunlight into food? Does the mesquite have a life of sensation or is it only in the meeting of tree and mind that feelings occur?

I feel I am inching my way toward Emerson's way of seeing, though I know I cannot get fully there. My sense of the divine is too compromised by data and doubt. I do know that the energy of emergence moves through me as it moves through trees, and this energy is the godliness of that imperative to form that shapes the universe, tree, flesh, mind, and this page. And all of it as an astonishment to hold dear and pass on down the line.

Growing through Fire

Maulian Dana

I REMEMBER SITTING IN MY FIRST SWEAT LODGE AS A teenager. Being from the Penobscot Nation and growing up around my culture, I was no stranger to the ceremony. I understood the protocols and mindset needed. I knew it was medicine not to be dishonored and not to be done by those who are not of a Native tribe. These are our teachings, and I was ready to learn them.

I am a panicky person, so when it felt hot and all the light was blocked in the lodge, my pulse throbbed and my hands tingled. The voice in my head that controls the anxious responses was screaming at me to bolt from the lodge, to climb over the others who sat in prayer, to get to the sweet cool air on the other side of the heavy blanket door.

I told myself to give it one more minute. To listen to the ancestors. I breathed in the fir bough and cedar steam, and although I didn't feel like I was breathing at all because of the thick heat, I felt a calming. A rhythm. The basic teaching of the sweat lodge is that you are returning to the womb. You sit in the heat of the steam from the rocks and feel the shifts in your creation. Mother Earth allows you to return to her, and when you pray in the lodge you are speaking to her.

The ceremony is a rebirth, which can push us and make us feel things we don't want to feel. Having given birth twice now I can understand the pain and also the beauty of being absolutely torn apart only to be greeted on the other side with a precious and perfect being. As mothers we bleed for the world. We give life by suffering and surrendering to the cycles of nature. We place our hopes for all of creation into the water, air, and earth as we channel our energy and power into our most sacred responsibility. We are reborn when we transform and grow through the fire.

The sweat lodge taught me how to be a mother because it brought me face-to-face with the parts of myself that needed to learn and suffer in order to be worthy and closer to my mother, the Earth. She is disrespected and demoralized and yet she still loves and cares for us. When I am faced with uncertainty and doubt in my journey that causes fear, I think about the strength and love that the Earth has for us even when we don't deserve it. I love my children without conditions, without fear, without thinking of myself. Mother Earth has shown me how. She models a love that is radical, real, and pure.

While it was part of my story, we don't all need a sweat lodge to know how to love our mother. We need to unite and love her while she is still here for us to appreciate. Without the mother there are no children.

Earth Verse

Kim Stafford

I WORK IN A SCHOOL WHERE INSTITUTIONAL WORDS sting my heart—words with rational intent but stressful effect: *due date, quiz, final, assessment, policy, security code. The agenda for the committee meeting.*

I feel better when we speak earth words: *dew, bud, rain, bee, wind, path, swale, river, moon.*

Even earth words of hardship or danger (*thorn, fang, cold*) don't pierce me as much as *digital template* or *strategic plan.*

Our days bristle with words that wound: *war, terrorist, extinction.* And school itself feels militarized when we *gather our forces, step into a minefield, attack the problem.*

Robert Macfarlane has noted words cut from the *Oxford Junior Dictionary*, including *acorn, buttercup, hazel, heron, nectar, otter,* and *kingfisher.* These had to go, editors explained, to make room for the likes of *attachment, blog, bullet point, celebrity,* and *voicemail.*

We engage in war against the earth—digging, drilling, plowing, polluting, "developing." Our behavior bruises Eden everywhere, but our unearthly language bruises our humanity. By contrast, when I scan a dictionary of the language of Hawaii, I see my poverty, for that language isn't limited to *rain, drizzle, downpour.* Islanders speak myriad affectionate words for *rain,* translated as *rain* (adornment of the gods), *fine light rain* (much loved), *bitter rain of grief, rainbow-hued rain, light-moving rain,* and *lunar rainbow* ("anuenue kau po").

The specificity of such language empowers more detailed seeing of particulars in the dense web of the wild. Without such precise words, landscape dwindles to a blur of vague beauty viewed askance through frosted glass.

In the *Tao Te Ching*, we find the ways of nature detailed as guidance for human behavior: "Plants in life are supple, in death withered. So tenderness is the attribute of life, stiffness the attribute of death." And in Old English, in passages from the primordial "Gnomic Verses," we find a similar recognition that humans will not thrive without an apprehension of the ways of rivers, ice, trout turning in a pool, the seasons of weather and of mood: *Forst sceal freosan, fyr wudu meltan* (Frost shall freeze, fire melt wood). Implication: a cold mood locks human exchange, but warmth may release us.

When we say something is *true*, we know the root of that word has kinship with *tree*, and also with *truce*. An authentic life can be schooled by the steadfast character and supple spirit of a tree. Knowing this, we may begin to make peace after our long war with fragile earth.

As a tree weaves a basket, as high tide knits debris along the shore, as the moon pencils shadows, we might each craft some simple form of earth verse. Drawn from the vocabulary of particular places in the landscapes most resonant for us and informed by traditions of observation and blessing, we might write by the trance of our own particular places in order to dwell in an earth-imbued garden of perception. To restore our buoyancy daily, we might gather the vocabulary of our natural places and distill a comforting spell.

Adapted from the introduction to *Earth Verse: Poems for the Earth*, by Kim Stafford (Little Infinities, 2019).

Visiting the Elders: Rocky Mountain Bristlecone Pine

David Haskell

BRISTLECONE PINE TREES LIVE ON THE EDGE OF POSSIBILITY. Their twisted trunks are sentinels at a boundary. On the slopes above them, only mosses and tiny flowers appear, on scree fields of bare rock.

My lungs grasp at the air, clutching at meager oxygen. It is midsummer on Mount Goliath in Colorado, yet the temperature is only a few degrees above freezing. When the wind comes, it slams me so hard that I stumble. The human body is ill-adapted to such extremes.

Yet "extreme" is a subjective judgment. For the bristlecone pines, this is home. They have taken into themselves the rhythms and substance of the mountain. Their limbs do not stretch out as other pines do, in long, simple curves. Instead, every branch and twig is a contortion, a twisting memory of the tree's conversations with wind and snow. The trees lean away from the prevailing wind, their branches trailing like flags.

The geometries of wood are especially evident where lightning strikes have slashed and torn away bark. Inside the trunks, fibers turn and wrap, a dance made of strands of wood. Tree resin colors these swirling patterns with varied hues of amber and teak. The sun has bleached the outer wood to gray. In the fresher wounds, places where lightning has recently touched, dark furrows mark the passage of the bolt of sky-power.

The wind has also left its mark in the material properties of needles and branches. The finger-length needles, heavily waxed to protect them from sun and wind, are bundled in fives. Each bundle is as stiff as wire. When the wind courses through the unyielding tines, the air is torn into a hiss. In contrast to the inflexible needles, every twig and small branch is springy, as if made of rubber. Many generations of heavy snow loads and strong winds have, through evolution, endowed the trees with this combination of tough needles and bouncy branches.

Fallen trees last for millennia here. The mountaintop is frozen for most of the year and, in summer, strong sun and wind dry the vegetation. Fungi and bacteria cannot prosper. By counting rings inside the wood of both living and dead trees, we know that the oldest tree alive in the Colorado bristlecone forests is 2,100 years old. A few others have lived for a thousand years. "Younger" trees date from the seventeenth and eighteenth centuries.

Rings from the 1830s and 1840s show deformations caused by cold winters and late springtime freezes. These years, rampant with typhoid and cholera, also killed many bison and people. The trees do not forget—they carry these winters inside their wood. The rings also tell the story of regeneration after fire. In 1625, 1700, and 1900, fires swept through this stand of trees, causing synchronized bursts of germination and growth.

How do the trees live so long? Unlike a tropical rainforest or a moist temperate woodland, the bristlecone pine grows largely unmolested by fungal enemies. But the slowness of decay is only part of the answer. The pine's internal rhythms also allow it to live long. Every needle lasts for fifteen years. A sapling may take a century to grow into a small woody stem. The leisurely tempo of the bristlecone's growth and physiology is an adaptation to the short growing season. The ground is ice-free and the air above freezing for only six to eight weeks every year.

The bristlecone pine tree lives "a long time." But perhaps these are the wrong words. The trees live not a long time but in a different time. Every creature has its rhythm.

The wildflowers that bloom between the trees and rocks grow only one or two centimeters above the ground. Yet the roots of these miniatures penetrate the gaps between rocks to a depth of a meter or more. What appears short-lived and fragile is in fact decades or centuries old.

A white-crowned sparrow sings from the top of the pine then flies to a willow thicket. My nerves are too slow to grasp the inflections of its song. The bird's nerves live in a different time too, fueled by a high body temperature. The microbes on our skin and in the soil around the trees likewise exist in their own time. Dozens of microbial generations may pass in a single human day.

The bristlecone pine's roots snake into rocks formed by magma that flowed to the Earth's surface 1.4 billion years ago. The mountain's gray rock is a monument to the incomprehensibility of geologic time. Ten billion new moons, a billion winters, a quadrillion cellular divisions made this mountain.

At any one place in the world thousands, perhaps millions, of times coexist. The land calls us out of our own time, drawing the imagination into a tempo incommensurate with our own.

Excerpted from an essay originally published in Italian,
"Il pino dai coni setolosi. L'albero che sfida clima e tempo," *La Lettura*
v 43 (October 28, 2018), 14–16. Translator, Maria Sepa.

Glory in Living Through Nature's Indifference

Juan Michael Porter II

WHENEVER I VENTURE INTO THE OUTDOORS, I AM reminded of my guileless passion for splashing through creeks and climbing over rocks without concern for my class or appearance. It is a freedom that I could never approximate in the concrete jungles of New York City, governed as they are by strictures of decorum and race.

But in nature, I am one with every tree branch or idle pebble. These creations are indifferent to and respectful of me, whether I successfully sail or falter over challenges. If I can maintain my balance on a boulder, then it will bear my weight. When I slip while passing through a ford, the water neither comforts me nor impedes my journey. Rather, it waits sans ulterior motive, as equally eager for me to continue charging forward as it is for me to dissolve into its stream.

Even this possibility of demise encourages me. When calamity ensues, the trees will not intercede to tip the scale in either direction. I am regarded as no greater than the pine cones or bees. It is between me and luck to determine what will happen. This is the epitome of freedom: living in a world that neither needs nor cares for me but that welcomes me all the same to accomplish whatever I can, if I can.

I relish this anonymity that the open air affords me. Because even as it has no consideration for my wants or history, it manages to alleviate the burdens that await me back home. Nature sees me as I am and welcomes me all the same, regardless of how others categorize me.

That invitation to breathe freely within a field of unspoiled beauty is constantly challenged by those who refuse to see me beyond the fear that they project onto

Black men. Their society has taught them that I will deliver great harm for no other reason than that I exist. These people, who I meet while rounding a treacherous ledge or stopping to rest after ascending a steep incline, tell me that they did not expect to find me within their pastoral retreat, despite my equal access to the same salvation.

I have no place within their grand vision of the boundless outdoors.

It is after these suffocating encounters that I am most grateful to the natural order. Nature lays no stake for either party. Rather, it pierces the choking miasma of racism with a pathway to brilliant vistas for anyone brave enough to abandon their suffering. The same offer exists for my antagonizers. Whether they take it or not is between them and their god; nature will wager no sides.

My challenge is to see beyond the invisible shackles that others attempt to place upon my neck; to embrace that it is enough that the outdoors can see me even when others will not. Nature cannot protect me, but nor will it deny me my divine right to its bounty. And so I seek new summits to surmount, year after year, knowing that whatever happens, existing as I truly am through nature's healing indifference is a glorious peace that I cannot bear to surrender.

On Being a Raining Cloud: From Rumi to Emerson

Alireza Taghdarreh

I WAS A CHILD WHEN MY FATHER GAVE ME A BOOK OF Rumi's poetry, called *Masnavi*, with its many Emersonian thoughts: "Be not like the water in the rain gutter, be the sky, be the cloud and rain." Rumi paved the way toward American transcendentalism. My enthusiasm for translating Emerson's *Nature* to Farsi emerged out of my love of Rumi and was so great that I spent more than a year on the careful study and analysis of the text. At one point, the original 1836 manuscript seemed so difficult to me that I shamefully decided to surrender. It was then that Emerson scholar Robert D. Richardson encouraged me to go on. He told me that Ralph Waldo Emerson is the best person to connect the cultures of the United States and Iran.

In a world where some call blind imitation virtue, and where the individual is sacrificed to consumerism, religious bigotry, and propaganda, Emerson's *Nature* can still save the souls of his new readers. His words invite them to see anew through their own eyes and free themselves from what others dictate.

Solitude, in Emerson's *Nature*, is the necessary condition for embracing the universe: "But if a man would be alone, let him look at the stars. The rays that come from those heavenly worlds, will separate between him and what he touches." By separating his readers from what touches and restricts them, Emerson extends their horizons and connects them to an ever-expanding world in which language, tradition, religion, and politics do not divide them from each other. On the journey from the Rumi of my grandfather to American transcendentalism, I have noticed many similarities between the cultures of Iran and of the United States. Rumi expands the world of his individual in a similar way to Emerson when he says, "You are not a drop in an ocean; you are an ocean in a drop." Rumi approaches the ocean the way Emerson moves toward the stars in his solitude.

Galaxies are nothing but a collection of single stars. In his starry solitude, Emerson finds reasons for unity among humanity. He shows us that the roots of our languages are in nature, which provides for our vocabulary, metaphors, and similes.

He invites his readers to transcend transactional religious doctrines emphasizing prize and punishment. In the Abrahamic religions, Adam and Eve are believed to have been expelled from Paradise for their sin. The rigorous goal of many followers of these religions is to repent, and so return to that paradise. But like many Persian mystics, Emerson believes that we should be the creators of our own paradise: "Every spirit builds itself a house; and beyond its house a world; and beyond its world, a heaven." Paradise is alluring, but why not fly higher?

In Emerson's dream world, we are a galaxy of individual stars sharing our lights together. In Rumi's, we are both a drop and an ocean. It is in this poetical empyrean that the peoples of the world can meet. After reading *Nature*, Emerson's readers will find they share their solitude with a broader, more expanded universe.

—For Wallace Kaufman and Robert D. Richardson,
mentors who turn seeds into stars

Roll and Hiss and Foam

Betsy Sholl

I HAVE OFTEN THOUGHT OF A PERSON'S CHILDHOOD landscape as a kind of inner geography, a psychic map that gives shape to the world and becomes an emotional and spiritual source. It could be any landscape, the concrete canyons of New York or the Appalachian Mountains. For me it's the New Jersey shore where I grew up in the 1950s and 1960s, two miles inland. River, marsh, bay, and then across the Mantoloking Bridge to the glitter of open sea—land touching water everywhere, "solid marrying liquid," as Walt Whitman says, boundaries and rigidities softening. My sisters and I stood in the surf as waves pulled the sand out from under our heels. We watched the tease of rippling horizon, where sky and sea seem to merge, each part of the other. We'd walk the long beach, heading into late-afternoon sun as if we could leave ourselves behind and enter that radiant light.

Of course, other realities existed. On the public beach the crowds were so thick in the summer that we couldn't walk to the water without kicking sand onto someone else's blanket. And beachfront property being private and filled with no-trespassing signs, we locals felt pushed out, resentful. But the shore up to the high-tide line was public, so there in the waves we belonged and were free. Always on that edge we stood with worldly concern at our back and looked out to that vast mysterious open horizon. The shore showed us our limits and gave us a vision of what is limitless, other. Now, of course, it also shows us how we have trespassed those limits. What seemed romantic as a kid—walking the rubble line, finding long coils of rope or other signs of distant shipwreck—now reminds us how endangered our oceans are.

The tides have a way of making what's broken beautiful—nautilus shells revealing their inner spirals, purple chips from quahog shells, sea-salted glass, and people as well. Once my high school boyfriend took me to meet his friend from the city, where he had once lived. The young man's girlfriend had gotten pregnant and had a baby at sixteen. At that time, shame and rejection were a common parental response. But these parents instead offered to support the couple so they could get their GEDs. In celebration, they all came to the shore: two teenagers with a baby, their four parents fussing lotion onto each other's backs, asking for the baby's sun hat, pulling out sandwiches and grapes and soda. My friend, whose own home life was troubled, said to me, "Wouldn't you give anything for that?"

Of course, it wasn't the beach that made those parents generous and practical, loving their children beyond disappointment into new life. But it was the beach that brought them to visit and allowed us to witness that kind of generosity. It is the shore that has shaped my spiritual life. What I long for spiritually comes from this geography: the ability to walk away from human distraction and pettiness and surrender to the utter unknown, that larger life force I'd call divine love, always wanting to draw us into its largesse.

Cold Water

William Powers

ONE DAY IN MARCH, I DRIVE TO A POND NOT FAR FROM my house. Crunching barefoot through the narrow apron of ice at the shore, I undress and dive in. I know this pond well from warmer months, but I'm not sure why I've taken myself here to swim on this particular frigid, late-winter day. I always scrupulously count my strokes, vestige of a childhood battle with obsessive-compulsive impulses. Now I swim until my body goes into panic mode: just ten strokes.

Strange behavior, in a way. I'm used to swimming in an indoor pool that is now closed due to the coronavirus pandemic. This is nothing like those toasty laps. But I keep going back. When the chill winds are blowing up whitecaps, I really have to steel myself to plunge in.

My stroke count gradually increases—30, 50, 100, 200. As I venture farther from shore, the pond gets quite deep and, when I open my eyes and look down, very dark. At a certain point, I feel the cold squeezing in on me from all sides and the survival alarms go off in my brain. I vamoose back to shore and scramble into my clothes.

I begin to notice something I've never felt in a warm-water swim: a sense of personal erasure or reduction. It's not so much the body that moves through life in these days of disease and quarantine; it's the mind, racing around the electronic networks and virtual platforms where we spend so much of our time. Maybe that is what makes me swim—and what I swim away from, thrashing for the void.

The mind often feels more like an adversary than a friend, doesn't it? That I have to work physically to escape it, plowing my way through this severe medium with only my limbs and lungs, must be part of the draw. No coincidence that my ritual began just as a global crisis was filling the days, even in my small town, with worry and fear. But I'd been craving this long before. Being locked down at home has given me the time and freedom to pursue it in earnest.

By late May, two things have happened. The water has warmed up, removing the need for self-encouragement and resolve. And other swimmers appear. Watching them slip into the pond with such ease and grace—water can transform even the most ungainly people into balletic sylphs—gives me an uncanny sensation of witnessing a small metamorphosis. Joining them feels, in one sense, like coming home. The virus is receding here, spring is here, and the element that has long been so forbidding and difficult, the water, is revealed as an old friend.

But there is also a sense of loss. The destination I've been approaching all these weeks yet never quite reached, the oblivion that was always about fifty arctic strokes away, is no longer out there. Oblivion is an odd goal, I guess, and even odder to have been chasing it for so long, punishing myself to try and get there. But I miss it.

The Geography of Memory

Akiko Busch

COMING INTO A SENSE OF PLACE HAPPENS OVER TIME, a process of acquisition that arrives in bits and pieces. I am reminded of this as I begin to compile a Natural Resources Inventory, a document that catalogs what we have—geology and soils, land cover, water resources, wildlife habitats—here in our Hudson Valley township in New York State. It is meant to be used for open-space planning, along with informing other various town plans and policies. The inventory will include maps of aquifers and floodplains, agricultural districts, stream habitats, connecting forests used by migrating species, as well as such landmarks as an old stone quarry and the fire tower perched on a ridgeline.

I've downloaded the GIS base map. Its information about land and space captures patterns and relationships between geographic features and land use. I see from the map, for example, that a section of road where I often find myself slowing down for deer, twice for a bear, once even a bobcat, winds through a wildlife corridor connecting two heavily forested areas. But the map needs updates and revisions. A trail to the fire tower on the summit of Clove Mountain was once a public hiking path valued by local residents; now it is private, its access restricted when a cell tower was installed there a few years back. Using my cursor, I trace a dotted area indicating a wetland, but I am sure that the marsh is larger than what the map tells me. Several roads through our own valley need identification. And topographical shifts indicated in shading reveal a rise I did not know of before. I move the cursor to a small blue patch indicating a stream-fed pond that should be just down the road from where I live, but the dam gave way last June, transforming the waterway into a thicket of shrubs and grass bisected by a thin creek. Still, there may be good sense in keeping the pond's historic Dutch name as it evokes bygone times of this small valley. Perhaps Vreed Pond will now be called Vreed Meadow.

I know that the layering of data on the base map echoes the way place is inscribed in memory over time. The more ephemeral geography of human recall relies on its own layering, its own varied typography, the imagery and lettering that form our sense of place. From the day we moved here more than thirty years ago, a towering sycamore at the edge of a hayfield down the road has been imprinted in my mind in boldface, and the name of Pray Pond where I swam one summer is written in lighter italics.

But other names are scribbled in more improvisational graffiti—a dirt road over the mountain that is closed off during the winter months; an old stone wall that scrolls through woodlands that were once pastures; a teetering silo that surrendered to gravity and weather a decade ago; and a forest of white pines where a palace of twigs, sticks, and leaves in the uppermost branches signals a restored eagle population. Even right outside my kitchen door, a slight cavity in the ground signals the place a massive swamp maple once stood. Diseased, riddled by fungus, and toppled more than twenty years ago, its buried roots have rotted, causing the dirt just above them to sink.

I know that the value of these maps, digital and cognitive alike, may lie in the fact that they serve as timelines too. In his book *Landmarks*, the British writer Robert Macfarlane observes the four ways in which landscape is lost: through the loss of beauty, the loss of freedom, the loss of wildlife and vegetation, and the loss of meaning. The last may be the most elusive, perhaps because how we find meaning in place has to do with those fugitive protocols that allow us to take measure of days, months, years.

And I think back to the woodlands that were pastures half a century ago, to the meadow that was a pond just last June, to a depression in the ground where a maple once stood. We humans are hardwired to find our place in things. Creating precise and efficient navigation systems is essential to our being, whether it is running a canoe with the current, using celestial navigation, or following a topographical map or satellite imagery. Yet at a time when we are likely to situate ourselves as dots blinking with pulsing immediacy in the center of a small screen, it may be useful to remember that how we find our bearings has as much to do with what once was as with what remains today.

Night Flight

Kimberly Ridley

IT'S BEDTIME, TEN O'CLOCK, BUT I CAN'T SLEEP. Something is tugging me outside on this cool autumn night. I get out of bed, grab a down sleeping bag, and settle into the hammock in the backyard.

Jupiter gleams over the silhouette of a spruce tree at edge of the field. Constellations shimmer and the Milky Way spills across the sky. Aspen leaves patter in the breeze, and then everything falls quiet.

I have spent many warm summer evenings in the hammock, listening to gray tree frogs and a Swainson's thrush sing in the dusk and watching fireflies until the mosquitoes drove me inside. I would welcome those mosquitoes now, anything to have summer back: the long, bright days, peonies in bloom, our woods and fields full of birds. I dread the long months of deepening cold and dark, the leafless trees, snow-slicked roads, the interminable Maine winter.

"*Tzeeep?*"

A small, high sound nicks the silence. I sit up in my sleeping bag.

"*Tzeeep?*" There it is again. "*Tzeeep...........Chip... Tzeeep.*" The sounds are coming from overhead. Flight calls. On this calm, clear night, songbirds are migrating, uttering these calls to keep tabs on each other in the dark. I can't see the birds or tell their flight calls apart, but I know they're up there: sparrows and thrushes, grosbeaks and buntings, vireos and warblers. Each species has its own distinctive call, most of which are beyond the range of human hearing. As hard as I listen, I hear only a tiny fraction of the birds aloft.

Unlike songs, which are lush and loud and all about territory and sex, flight calls are terse and to the point. I imagine the birds' nocturnal conversations:

"You there?"

"Yup."

"Where?"

"Over here"

"Okay."

I wish I could see all those birds above me, flashy songbirds like indigo buntings and Baltimore orioles, scores of streaky brown song sparrows, and dozens of jewel-toned warblers—northern parulas, black-throated greens, magnolias, and all the rest. I shine a flashlight into the sky. A few small, white moths flash in the beam, but no birds. They're too high, flying between one and two thousand feet over the Earth. What I really wish is that I could fly to the tropics with them.

That's not an option, so I lie here, earthbound, gazing up at the sky. I think about all those birds in the dark, their wings beating twenty times per second, their tiny hearts and lungs pumping as they gulp about a hundred breaths a minute.

It surprised me to learn that most songbirds migrate at night, but it makes sense for several reasons. The air is usually calmer, cooler, and more humid at night, which helps prevent overheating and dehydration as birds make their arduous journeys. Traveling at night also helps birds avoid predators such as hawks, which migrate during the day.

Even more astonishing: birds use the movement of the stars as one of several navigational aids. In experiments conducted in planetariums, scientists have shown that migratory birds orient themselves south in the fall and north in the spring using the rotation of stars around a "fixed" point in the sky, which is Polaris, or the North Star. When scientists moved Polaris in a planetarium's ersatz sky, or used a different star as the point around which the constellations wheeled, the birds reoriented themselves accordingly.

"*Peeeep!*" A delicate whistle pierces the night. This one I know: a Swainson's thrush, whose flight call sounds like a spring peeper. It's probably not the thrush I listened to this summer but a traveler from farther north.

Twigs snap. Bushes rustle at the edge of the lawn. My heart races, along with my imagination. Utterly still in my down cocoon, I peer into the dark, my nerves flirting with the unseen and unknown. In a few minutes, the rustling fades away into the woods. Deer perhaps, or raccoons, wild neighbors on their evening rounds.

A meteor streaks across the sky. The constellations shift. It's getting late and I should go back to bed, but not just yet. I hunker into the warmth of my sleeping bag. The stars burn and burn, an invisible river of birds flowing below them.

Originally published as "An Invisible River of Birds"
in *Christian Science Monitor*, October 14, 2015.

Breathless

Paul Bennett

THE FIRST FIFTEEN FEET FLURRY. WATER SWIRLS. My fin slaps the surface with a crack. As the pressure increases in my ears, seemingly audibly, I use my hands for a boost. But it's the kick that does most of the work as I move downward into the depths.

Yellow fusiliers, chubs, and scads, each in their distinct shoals, school suddenly as I dive down. A concussive whoosh, like someone shaking out a wet sheet, passes through the water as I move between them—a solid, moving wall of fish shimmering in the sunlight that splits and then surrounds me. A few seconds later they disappear above.

At sixteen feet the empty blue takes over. Fishless and featureless, there's only pressure—nearly half an atmosphere's worth of water above me. It screws my mask tight. My ears beg forgiveness. The physics informs the biological as my heart rate palpably slows. I kick my fins slowly now but powerfully. It's impossible to judge the speed of the descent without any frame of reference. Just me in the blue.

Our experience of the marine environment usually has a focal point—the waves on a beach, a reef, a fish, turtle, or mammal. But the sea offers no inherent focal point. I've learned this by sailing more than twenty thousand nautical miles with my family, gazing out at the endless ocean on passages across the Pacific. Each swell may have an individual character, but this is subsumed and absorbed by the inchoate oneness of the sea—angry, green-blue, blank, godlike.

Which is maybe why I've taken up free diving. Unencumbered by heavy gear, free diving promises complete immersion in the otherworldly. It brings into relief the invisible hurricane of physical laws that surrounds us and seems to slow time itself.

At twenty-six feet the two pounds of lead around my waist takes over. I am negatively buoyant now and sinking and can stop kicking, saving precious energy. My alveoli, the lung's tiny air sacs, are less desperate. The reef comes into view—a blob of yellow at first, slowly sharpening. A smooth, narrow ridge of coral rises up to a point and travels in a line as far as I can see. I've been down for just twenty seconds so far.

Pungu swim wide-eyed up from the bottom, curious of my arrival. Masses of them. If I had my speargun I might catch one. They taste excellent, especially when we grill them on the stern of the boat with a little thyme. But, I have no gun. Besides, the reef calls me deeper.

At almost fifty feet, I arrive. With my gloved left hand I gingerly grasp a round knob of coral that sticks out from the wall, helping me hover horizontally in space. Ahead, green, yellow, and blue corals interlace in a non-Euclidean landscape that bends into fantastic, otherworldly shapes. A sea fan arches across an opening. A pinnacle of *Favites flexuosa*, a round dome of coral etched with pentagonal ridges as if commissioned by architect Lord Norman Foster, stands solitary on a platform. Fish flash through the geometry in bright colors: wrasses, butterfly fish, a coral grouper with its massive red head and psychedelic blue dots as if it's headed to a party. There is a flat ledge just beneath me that extends a short distance and then falls off into emptiness.

I've been underwater for a minute now and feel my chemistry straining: dissolved oxygen pressing through narrow vessels like metro passengers struggling to board a crowded train. But I know that I've got another thirty seconds, at least, if I ignore it. It's a mind game now. Less physical and more psychological.

This underwater world is incredibly loud, rather than deafeningly silent as some believe: Corals pop, parrotfish crunch. The pressure in my ears thrums with a constant monotone. And everywhere movement. The reef stands in opposition with the blue beyond: It's urban pandemonium. An underwater Jakarta or Lower East Side.

A school of fifty barracuda turn sharply toward me and then sharply again to cruise along my ledge. They are checking me out, this foreign object, and grinning through their toothy underbites. Then, a flash, as a dog-tooth tuna and his mate glide through them. They, too, are hunting and keep a wary distance.

I could get lost in this. All about me the reef swirls with cacophony while inside is a pressurized calm. My mind, alert to each movement and sound, accompanies an inner deoxygenated *om*, keeping me still. The balance is powerful, and sometimes I feel like I could just let go and stay forever.

But, at bottom, we're all physics. I need air. I feel it in the tightening of my chest and frisson in my spine. My watch, which reads 1:25 minutes, underscores it. I let go, rotate upright, and begin a light, swaying kick upward. The reef recedes as I reenter the blue.

A gray reef shark emerges from the blank beyond and swims in an *S* toward me. He, too, is curious. Where did this creature come from, and is it meant for me? He circles wide, eyeing me, then turning away and downward. My eyes follow as the distance between us widens. I wonder about him: Will he simply wander the empty blue, embalmed in the featureless open of the sea? Or is he headed to the reef? I envy that he can remain in this limbo that I only partially experience, one breath at a time.

Written aboard *Dafne*, anchored off Sumatra, Indonesia.

The Indispensable Oak

Doug Tallamy

WHEN I WAS TEN YEARS OLD, I BOLDLY DECLARED THE white oak to be the very best tree there was. It certainly was the best tree I had ever experienced. I don't remember exactly why I made my declaration. Perhaps it was because I liked superlatives: after all, oaks were the largest and oldest trees I had yet encountered, and they were the best climbing trees, especially when grown in an open field. I also just liked the way they looked. Little did I know that fifty years later, research from many fronts, including my own, would show that my young self had been quite right; oaks are indeed superior trees, and the future of conservation efforts across the United States will depend on them.

If we are to develop a sustainable relationship with the natural world—and since it is nature that sustains us, having an unsustainable relationship with nature is simply not an option—then landscapes everywhere, including those in which we live, work, and play, must do four things.

First and foremost, human-dominated landscapes must support complex and stable food webs. The plants we put in our landscapes must pass the energy they harness from the sun on to other living things, or there will be no other living things. No plant does this better than oaks. Across North America, oaks feed nearly a thousand species of moths and butterflies, far more than any other plant group. Through the insect populations they support, oaks are a major source of food for dozens of bird species, and most birds need to consume hundreds of insects every day. Oaks support other invertebrates as well, including cynipid gall wasps, June beetles, longhorned beetles, metallic wood-boring beetles, weevils, myriad spiders, and dozens more species of arthropods, mollusks, and annelids that depend on oak leaf litter for nourishment and protection. Before it dies, a single tree can also drop up to three million acorns, serving as a lifeline for squirrels, chipmunks, and other rodents; bears, deer, racoons, and possums; and also birds like jays, towhees, titmice, red-bellied woodpeckers, turkeys, and wood ducks.

All landscapes must also sequester carbon, in the plant tissues they house as well as in their soils. By virtue of the enormous size of both their above- and below-ground parts, oaks are one of our best plant choices for fighting climate change through carbon capture. They can live for hundreds of years. An oak tree is the ecological gift that keeps on giving. It will store carbon out of harm's way for a very long time.

Another indispensable ecosystem service all landscapes must provide is watershed management. There are few plants better than oak trees at reducing the impact of pounding rain and, through its roots, trunk, and leaves, at holding stormwater on site until it can percolate down to the water table.

And all landscapes must support diverse communities of pollinators. Until recently, I thought this was the one ecological skill oaks lacked because they are wind-pollinated. Yet several recent studies show that up to 80 percent of the pollen collected by some of our native bee species in the springtime is from oaks. Who knew?

I have planted ten species of oaks on our property, for sustainability and the fact that I just plain like them. Over the years I have photographed 1,028 moth species as they developed on our oaks, and because our oaks make so much "bird food," my wife, Cindy, and I have enjoyed the comings and goings of fifty-nine species of birds that breed here, and many more that stop to eat during their migration. Our oaks have entertained us daily for nearly twenty years. I cannot imagine life without them.

The Anti-Garden: Husbandry Versus Wildness

Jinny Blom

WHAT IS WILD?

Gardening exists on the luxurious fringe where the earth has been tamed and there is sufficient spare time to start embellishing it. As a landscape designer, it fascinates me that there is presently a fashion for "wild" gardening. Wilderness has captured the gardening imagination.

To garden is to create, to imagine, and to tinker, so professing it "wilderness" seems a bit counterintuitive. However much we wish to create the wild, it can only be a construct and is therefore, in my opinion, always a garden.

Making a wild garden still requires a lot of structure. At one site, the owner, a free-thinking academic, wanted to subvert a perfectly logical garden that surrounded her Tudor farmhouse in the country into something altogether wilder. She wanted her children to enjoy the experience of boundless nature rather than acres of manicured lawn. We came up with the idea of an "anti-garden."

This act of subversion required thought. How far were we prepared to go in "letting go" the land? I cautioned that it would be very hard to wrest it back from nature.

Inspired, we let the lawns grow mostly unchecked but planted an Orwellian vegetable patch on the front lawn to ensure that radishes were easily accessible. We let the house become submerged in roses and vines and creepers but turned the rest over to masses of rambling orchards and nutteries. The gardener went pale.

This layout required a strong underlying design to hold the whole rationale together. I planted wild nutteries; managed nutteries, orchards, coppices, and a border of herbs; and smothered the house and outbuildings completely in Rosa 'Madame Alfred Carrière', Rosa 'Rambling Rector', Rosa *banksiae* 'Lutea', Virginia creeper, *Vitis*

'Fragola', and *Vitis vinifera* 'Purpurea', with oxeye daisies everywhere. It was so exciting. There was such an abundance—of bees, butterflies, birds, and slow worms.

The anti-garden has stabilized nicely into a simple system of management. Perhaps the greatest successes were the reverted lawns and the submerged house and the array of wildlife that is now at home there.

Adapted from an essay originally published in
The Thoughtful Gardener: An Intelligent Approach to Garden Design,
by Jinny Blom (Jacqui Small, 2017).

Ecosystems of Friendship and Water

Thomas L. Woltz

IAN QUATE AND I MET AS CLEANUP VOLUNTEERS AT Brooklyn's Gowanus Canal, one of the most contaminated water bodies in the United States. Our first conversation revealed we had both grown up in the French Broad River watershed in western North Carolina. There began the aqueous theme of our friendship. Ian later proposed we collaborate to document the unique extremophiles living in the toxic stew of the Gowanus before it would be capped by the EPA. As landscape architects, we would go on to design projects to protect fens and wetlands and to improve our collective relationship within water systems by revealing and marking ancient flows—bringing us to the places where water springs forth from the earth and closer to the river's edge.

About the same time, I formed another aquatic-based friendship with Florida artist Margaret Tolbert. A poet, painter, and oracle of the springs, she invited me to swim in the sacred waters of Ocala National Forest. Her dazzling descriptions of the springs were peppered with references to ancient Turkey, Italy, and Egypt. I invited Ian, always game for adventure, to join me on a very indirect route to North Carolina.

Margaret drove us from the Gainesville airport directly to the first of eight springs we would visit in two days. Gilchrist Blue Springs is a white limestone basin amid a native hardwood hammock. The autumnal chill was dispelled by the constant 72-degree water. Borrowed flippers enabled diving against the surge of 44 million gallons daily to explore sparkling domed caverns and crevasses. I imagined the relative scale of our three bodies held aloft by the powerful current inside a limestone Pantheon of water, the oculus above sending shattered rays of light to the depths of the cavern. After more springs, sand boils, and a three-mile downstream swim in the Ichetucknee River, Margaret delivered us to the airport, fresh from the Devil's Eye Spring, the last of eight. We flew on to Asheville.

When Ian and I travel to North Carolina, our first discussion is choosing hikes. This trip, I proposed we start at my family's property adjacent to the Pisgah National Forest and hike the famed Cold Mountain. We followed the creek that my grandfather, father, siblings, and I had regularly played in and fished, then branched up the steeper incline to the 6,030-foot summit. Placing hands on the 480-million-year-old exposed bald granite, I connected with one of the most ancient places on Earth. The deep cracks between the smooth stones, like some pachyderm's skin, were filled with blueberry shrubs in scarlet fall color, like flowing lava. Near the summit a small pipe fitted deep into a massive granite boulder gushed spring water…the tiny headwater of the French Broad.

As I drank and the cold ribbon of water passed into me, the aqueous connectivity of the past twenty-four hours and the ecology of a friendship appeared clear and inseparable. The same spring water in which our two bodies were suspended hours earlier in Florida was in the ribbon of water surging upward through the mountains between sheets of granite and into our bodies. A fluid friendship entwined in an aquatic continuum from an ancient granite seep to the mysterious karst pools five hundred miles away.

Agricultural Humanity

Gene Baur

THE BUMPER STICKER READS, "HUMANS AREN'T THE only species on earth; we just act like it." This witty take highlights how our species' arrogance has been harmful to other animals, nature, and ourselves. Our sense of entitlement and failure to respect others is reflected in many ways, including in what, or who, we eat.

We slaughter billions of farm animals every year, treating them like unfeeling commodities. Our excessive consumption of meat, dairy, and eggs contributes to the destruction of diverse ecosystems around the globe, used for pasture and cropland to feed these animals. This undermines planetary resilience and biodiversity. It's a significant driver of the climate crisis, which threatens us and other species.

The food we ingest represents one of our most intimate connections with the earth, and our agricultural system helps define our relationship with nature. The routine abuse of other animals and the environment to extract a perceived financial or material benefit exemplifies our lack of respect.

Rather than squandering precious resources, we should seek to develop mutually beneficial relationships with the natural world, starting with our food system. We can feed more people with less land and fewer resources—while enhancing our own health and humanity—by eating plants instead of animals.

Reforming agriculture will require structural shifts in land use and government policies as well as cultural evolution. Throughout human history, we have been drawn to meat, just as we have been attracted to power. Societies with the greatest wealth and power have tended to consume the most meat and also to cause the most harm.

Scientists now warn that we are living in the Anthropocene, a geological epoch that will be marked by human dominance, species extinction, and a fossil record littered with plastic and chicken bones. But it doesn't have to be this way. By changing our food system, respecting animals and nature, and acting with greater humility, we can help other animals and ourselves.

In the United States, ten times more land is used for animal agriculture than for plant-based production. Shifting to plant-based farming would free up millions of acres to feed ourselves and also create space for diverse ecosystems and wildlife habitats that support clean air and water. Our farmland can be managed more responsibly, using organic and permaculture principles that encourage biodiversity, instead of the toxic petrochemical monocrops we grow today.

We can also grow food in urban and suburban settings. It's promising to see the spread of farmers' markets, community supported agriculture programs, and community gardens. Abandoned city lots, school yards, and suburban lawns are being reclaimed to grow healthy food, also creating meaningful jobs and opportunities in the process.

Our food system has profound consequences for ourselves, as it does for the other species we share this planet with. Albert Einstein said, "Our task must be to free ourselves from this prison by widening our circle of compassion to embrace all living creatures and the whole of nature in its beauty." We can liberate ourselves and help heal the planet by changing the way we eat.

Life Cycles

Wallace Kaufman

SEEING A GIRL-CHILD LOOKING SADLY AT THE FALLING leaves of a tree, the poet Gerard Manley Hopkins wrote:

> Ah! ás the heart grows older
> It will come to such sights colder
> By and by, nor spare a sigh
> Though worlds of wanwood leafmeal lie.
>
> *From "Spring and Fall"*

Hopkins was keenly aware not just of death in nature but of his own mortality. I often thought of these words as I cleared a small patch of hillside hardwoods for my house overlooking Morgan Branch in North Carolina, accompanied by my four-year-old daughter, Sylvan. The leaves that fall at the end of summer's shade eventually rise again for another summer.

Every day I went into the woods, I was aware of this great rising and falling and circling of life. As I pried blue stones out of the red earth, I watched Sylvan playing with rocks or studying an ants' nest or drawing pictures on scrap paper. At thirty-four years, I was already declining while she was rising. I knew that when she was my age she would remember almost nothing from these days. What did I remember? My grandmother singing the Tom Thumb nursery rhyme about the little man who killed the little duck. I remembered the green park across the street from her house and in it, live ducks. I told Sylvan as much as I thought interested her.

We often sat together and watched birds and frogs and snakes and crayfish and examined flowers and smelled the root beer smell of wild ginger. I knew she would not remember the days, the events, or any single animal or plant, yet I also knew that we seldom remember the first encounter with anything we are capable of loving deeply, anything that is most comfortable and familiar.

Among the things I showed Sylvan was also death. We found a dead fox lying by the creek, its white bones beginning to emerge through the rotting pelt. I explained to her what was happening. I showed her the beetles and maggots that were making their living from the fox. We poked through the life in rotting logs. I pried apart the owl pellets and showed her the pieces of bone, teeth, and hair of the mice and voles that had kept the owl and its babies alive.

I knew the education I was starting for Sylvan was as important as the statutes her mother was then studying in her second year at law school. Sylvan was in her first year of natural law school, and as important as law is to civilization, what she could learn at Morgan Branch were unchangeable laws. Neither her mother nor I attended church, and I knew Sylvan was growing up without the comfortable sense that beyond death lay Heaven. The best I could do was to be sure that she understood that life leads on to life.

Now, seventy-seven years later, I hope she will assure my final wish to be buried in a simple box and above me, ready to send its roots downward, a black walnut or the acorn of a white oak. It's the closest I'll come to resurrection and becoming a permanent resident.

My epitaph:

Here lies a man
Whose life was long
Whose will was weak
His body strong.

His first, his final hope
To feed the world's wonder
Growing ideas, harvesting words.
Now he feeds this tree he's under.

Nature's Seasons

Max Morningstar

BUNCHING FRENCH RADISHES IN THE FIELD, MY HANDS are engaged and working. During the summer, mosquitoes hover by my ears and neck. Helpless, unable to shoo them away, I am left anxious by their high-pitched whine. When the dragonflies arrive to hunt the mosquitoes at dusk, it is easy to believe that they have heard my silent cries for assistance. They flutter, dart, bob, and weave in a form both elegant and robotic. They hunt the mosquitoes and pluck them from the air around me, their wings brushing as they fly by.

I do not seek out nature. Farming is my livelihood. I live almost every day in a world of the soil, air, and wind. Farming for profit is an overwhelming job. I manage tasks that have to be completed with never enough time. During fourteen-hour days on the farm, I have the opportunity to catch nature at her most poignant and most subtle. The sound and sensation of the dragonfly is a quick buff, a gentle knock as the wings brush my ears and bring relief. Ah, that is one of my favorites.

The seasons are nature's clock and calendar, phases to live within and revolve around. I cannot push nor pull them, nor convince them to slow nor hasten. Nature is wildly beautiful and deeply cruel, and no matter how many times I have begged, she has moved without me, changed without me, with no notice. She is neither benevolent nor spiteful, but indifferent. We are a part of all, and she knows.

An eternity's worth of experience passes in a moment, only to be realized after it is gone. We are enamored and surrounded by our tasks, by the work we create, and then one day nature has changed again. Cooler, shorter days are the sign of a new season.

The barn swallows are gone, fleeing the great north wind.

The lightning bugs that romance our summer evenings have found a place to sleep.

The dragonflies, having feasted well, with their offspring left in ponds for spring, have passed.

And we remain. The ebbing and flowing slows, the color fades from green and brown to shades gray and white and black. We promise ourselves that we will notice more and appreciate more next time, before it has gone.

Teasel, A Healing Plant

Deb Soule

FOR MANY YEARS, TEASEL (*DIPSACUS SYLVESTRIS*) HAS been thriving in my healing garden at Avena Botanicals farm. The biennial herb prefers full sun and grows as a rosette the first year, its thorny leaves lying flat upon the earth. During the summer of the second year, teasel sends up a four- to five-foot-tall flower stalk. Elongated leaves clasp the sturdy stem, creating magical cups that catch and hold rainwater. Near the top of the branching flower stalk, several thistle-shaped heads appear. Then, over the next few weeks, tiny lavender flowers magically begin to encircle the flower heads, attracting native bumblebees and, on occasion, a ruby-throated hummingbird.

The first-year root, dug in autumn, is the part of the plant herbalists dry for tea or chop and prepare into a fresh root tincture, offering medicinal benefits for mending broken bones and helping people with Lyme disease.

Teasel's most magnificent gift is the water held within the vessel created by the leaves. This healing water restores brokenness of spirit and repairs holes in the energy field of a person or animal. It was from observing a mourning dove with a broken wing in my garden that I first learned about teasel's power to heal.

One morning in May while shoveling compost into a wheelbarrow, I noticed a mourning dove hovering near my compost pile for warmth. I gently lured her to the safety of my garden by scattering sunflower seeds for her to follow. For several weeks she lived in my garden, coming out from the protection of the hawthorn trees to sip the water held within the teasel's leaves. She sipped this healing water each day. I also noticed another mourning dove nearby and heard them communicating with one another. One day, she flew upward out of the garden with the other mourning dove. I saw them sitting together on the electric line. Tears of joy ran down my cheeks. Later that summer, baby mourning doves were born near the edge of my garden.

The mourning dove who came to my garden showed me the powerful medicinal qualities of teasel, honored in Chinese and Western herbal medicine. For many years now, I have suggested three drops of a teasel root tincture or a teasel flower essence for people with broken bones, for those whose spirit feels broken, or when a person feels emotionally exhausted, depleted, discouraged, vulnerable, or at odds with the world. Teasel helps the soul return to a place of inner harmony and peace.

The Tide

Kathleen Dean Moore and Erin Moore

LOOK, FIVE WHITE EGRETS ARE LINED UP ON THE dike, waiting. Five yellow bills slowly swivel, following the tide as the last of it slides off the mudflat. Strands of eelgrass point to where the tide has gone. At the near edge of the slough, sedges stick up like birthday candles waiting for the match, and salt marsh bulrushes shine with salt. A northern harrier soars over the emerging mudflat, adding to the *mute music* of the morning.

Since dawn, an eagle has perched in the beach pine at Lone Rock. Finally, he unfolds his wings and drops onto rising air. As if the movement were a maestro's downbeat, all the marsh shouts out. A flock of mallards and as many wigeon thunder into the air, and hidden herons flush, croaking. A salmon, or maybe it's a flounder, smacks into deeper water, leaving a whirlpool on the slough.

Then there is no hushing the birds. Maybe spring arrived at that exact moment, or maybe it was sun that warmed the mud to awakening, or maybe the final falling of the tide in one splendid moment melted the *frozen music* of the architecture of tussocks and mud-shrimp towers. Whatever it is, the varied thrushes will not stop whistling. A red-winged blackbird yodels from his perch on a surveyor's stake. A nuthatch toots his bent tin horn. The air above the mud begins to shiver, whether in warmth or fear, I do not know.

I come to a special viewpoint to watch morning tides like this. An architectural installation sited at the edge of the water, it is a wide, wooden cylinder constructed of circular trusses and thatched with sedges harvested in the slough. Tall enough to stand in, with a bench fitted nicely to my knees, the tunnel is maybe four paces long. It focuses my vision like a spyglass. It commands me to see.

See this. See this, the viewpoint demands. You may not turn your eye away. These birds, these beings, these small lives that cannot keep from singing?—these are the elders chained to the tracks in the way of the oil trains. These are the neighbors blocking a truck loaded with pipe. These are the handcuffed women and their bewildered children. These are the glorious beings who flutter in the way of the incoming tide of the liquid natural-gas terminal planned for this place: concrete pads, explosive gas, dredging spoils, bulldozed slash, inconceivable amounts of cash, government collusion, and a fleet of white F-350 diesel trucks. Lies, lies, lies. See these birds. See their glory and the sins plotted against them. *Nature thunders to man the laws of right and wrong*. It is right when a marsh at daybreak shimmers and sings; that is as it should be. Whatever destroys its joy, its beauty, the ancient urgency of its lives—that is wrong.

Caring for Nature

Rachel Carson

THE WORD *NATURE* HAS MANY AND VARIED CONNOTATIONS. I like this definition: "Nature is the part of the world that man did not make."

Humans have long talked somewhat arrogantly about the conquest of nature; now we have the power to achieve this boast. It is our misfortune—it may well be our final tragedy—that this power has not been tempered with wisdom, but has been marked by irresponsibility; that there is all too little awareness that man is *part* of nature, and that the price of conquest may well be the destruction of mankind.

Measured against the vast backdrop of geologic time, the whole era of mankind seems but a moment—but how portentous a moment! It was only within the past million years or so that humans arose.

Who could have foretold that this being, who walked upright and no longer lived in trees, who lurked in caves, hiding in fear from the great beasts who shared our world—who could have guessed that we would one day have in our hands the power to change the very nature of the earth—the power of life and death. Over so many of its creatures?

Who could have foretold that the brain that was developing behind those heavy brow ridges would allow humans to accomplish things no other creature had achieved—but would not at the same time endow wisdom so to control activities that would bring destruction upon ourselves?

Our attitude toward nature has changed with time. Primitive men, confronted with awesome forces of nature, reacted in fear of what they did not understand. They peopled the dark and brooding forests with supernatural beings. Looking out on the sea that extended to an unknown horizon, they imagined a dreadful brink lying beneath fog and gathering darkness; they pictured vast abysses waiting to suck the traveler down into a bottomless gulf.

Only a few centuries have passed since those pre-Columbian days, yet today our whole earth has become only another shore from which we look out across the dark ocean of space, uncertain what we shall find when we sail out among the stars, but like the Norsemen and the Polynesians of old, lured by the very challenge of the unknown.

Between the time of those early voyages into unknown seas and the present we can trace an enormous and fateful change. It is good that fear and superstition have largely been replaced by knowledge, but we would be on safer ground today if the knowledge had been accompanied by humility instead of arrogance.

Human's attitude toward nature is today critically important, simply because of the newfound power to destroy it. I clearly remember that in the days before Hiroshima I used to wonder whether nature—nature in the broadest context of the word—actually needed protection from man. Surely the sea was inviolate and forever beyond man's power to change it. Surely the vast cycles by which water is drawn up into the clouds to return again to the earth could never be touched. And just as surely the vast tides of life—the migrating birds—would continue to ebb and flow over the continents, marking the passage of the seasons.

But I was wrong. Even these things, that seemed to belong to the eternal verities, are not only threatened but have already felt the destroying hand of humans.

Water, perhaps our most precious natural resource, is used and misused at a reckless rate. Our streams are fouled with an invisible assortment of wastes—domestic, chemical, radioactive—so that our planet, though dominated by seas that envelop three-fourths of its surface, is rapidly becoming a thirsty world.

In 1955 a group of seventy scientists met at Princeton University to consider mankind's role in changing the face of the earth. They produced a volume of nearly 1,200 pages devoted to changes that range from the first use of fire to urban sprawl. The distinguishing feature of human activities is that that they have almost always been undertaken from the narrow viewpoint of short-range gain, without considering either their impact on the earth or their long-range effect upon ourselves.

The stream of time moves forward and humans move with it. Our generation must come to terms with the environment. Our generation must face realities instead of taking refuge in ignorance and evasion of truth. Ours is a grave and sobering responsibility, but it is also a shining opportunity. Go out into a world where mankind is challenged, as it has never been challenged before, to prove its maturity and its mastery—not of nature, but of itself. Therein lies our hope and our destiny.

Rachel Carson, speech, Scripps College Commencement, June 1962,
Rachel Carson Papers, Beinecke Library, Yale University.

Contributors

Gene Baur, hailed as "the conscience of the food movement" by *Time* magazine and named one of Oprah Winfrey's SuperSoul 100 Givers, cofounded Farm Sanctuary in 1986. A pioneer in undercover investigations, Baur was instrumental in passing the first US laws to prohibit inhumane animal confinement and continues to work on systemic food-industry reforms. Bestselling books include *Farm Sanctuary: Changing Hearts and Minds About Animals and Food* (2008) and *Living the Farm Sanctuary Life* (2015).

Writer and entrepreneur **Paul Bennett** lives on a sailboat in Indonesia with his wife and three daughters. He's written extensively for *National Geographic* and *Outside*, authored several books about design and architecture, and won a Lowell Thomas Award for Travel Writing. He cofounded and ran the expert-led travel company Context for many years.

After a first career as a psychologist, **Jinny Blom** chose to follow the thread of human health through landscaping, founding Jinny Blom Landscape Design. She believes we are a product of our physical environment and that landscape gardening is a synthesis of everything that matters.

Akiko Busch is the author of *How to Disappear: Notes on Invisibility in a Time of Transparency* (2019). Her previous essay collections include *Geography of Home* (1999), *The Uncommon Life of Common Objects* (2004), *Nine Ways to Cross a River* (2007), and *The Incidental Steward* (2013). She was a contributing editor at *Metropolis* magazine for twenty years, and she swims across the Hudson River at least once a year.

Rachel Carson (1907–1964) was an American marine biologist, author, and conservationist whose book *Silent Spring* (1962) expanded awareness of and advocacy for environmental conservation.

Maulian Dana serves as the Tribal Ambassador for the Penobscot Nation in Maine. She is a Tribal Citizen and mother of two who raises her children in the reservation community where she grew up. She holds a BA in political science and her role in the tribe is representing the Penobscot Nation to local, state, and federal governments as well as advocacy and policy work. She is also an avid reader, writer, and activist for racial equity.

Poet and essayist **Alison Hawthorne Deming** is author of the poetry collection *Stairway to Heaven* (2016) and the forthcoming work of environmental and cultural history, *A Woven World: On Fashion, Fishermen, & the Sardine Dress* (2021). She is Regents Professor at the University of Arizona and has been recipient of a Guggenheim Fellowship and Walt Whitman Award. She lives in Tucson, Arizona, and on Grand Manan Island, New Brunswick, Canada.

David Haskell is the author of prizewinning books *The Forest Unseen* (2012) and *The Songs of Trees* (2017). His work integrates scientific, literary, and contemplative studies of the natural world. He is professor of biology and environmental studies at the University of the South, Sewanee, Tennessee.

Wallace Kaufman's latest book, *Grow Old and Die Young* (2019), is an illustrated memoir. Having traveled and worked in many countries and wild places, such as northern Siberia, he makes his home between forest and tidal slough in coastal Oregon. Kaufman was a science writing fellow at Woods Hole Marine Biological Laboratory.

Stuart Kestenbaum, editor of *Visualizing Nature*, was the director of the Haystack Mountain School of Crafts in Deer Isle, Maine, for twenty-seven years, where he established innovative programs combining craft with writing and new technologies. He's serving a five-year term as Maine's poet laureate, through 2021. Kestenbaum is the author of five collections of poems, most recently *How to Start Over* (2019), and a collection of essays, *The View from Here* (2012).

Erin Moore is an architect and a faculty member in architecture and environmental studies at the University of Oregon. Moore uses her design practice, FLOAT architectural research and design, to explore ways that buildings shape and reflect cultural constructions of ideas of nature.

Kathleen Dean Moore is the author or coeditor of a dozen books about our moral and spiritual relation to the wild, reeling world. Formerly a distinguished professor of philosophy at

Oregon State University, she left her position to write about the moral urgency of climate action in books such as *Moral Ground* (2010) and *Great Tide Rising* (2017).

Organic farmer **Max Morningstar** is dedicated to growing and distributing high quality produce throughout the Hudson Valley, the Berkshires, and New York City. The founder of MXMorningstar Farm in Hudson, New York, Morningstar uses sustainable practices designed to maintain and build the health of the land for years to come.

Juan Michael Porter II is an arts and culture journalist dedicated to the intersection of Black lives, media criticism, and HIV advocacy. He is a National Critics Institute Fellow and a member of the American Theatre Critics Association and has written for TheBody, TheBodyPro, the *Washington Post*, *SYFY Wire*, *Observer*, *TDF Stages*, *Time Out New York*, *American Theatre* magazine, *Colorlines*, *AMC Outdoors* magazine, *Anti-Racism Daily*, *SNews*, *HuffPost*, *Broadway World*, and *Ballet Review*.

William Powers is a writer and technologist and the author of the *New York Times* bestseller *Hamlet's BlackBerry: Building a Good Life in the Digital Age* (2010). He has spent the last six years working on research projects in artificial intelligence at the MIT Media Lab and the Center for Humans and Machines in Berlin, where he is the visiting scholar for humanistic technologies. He lives on Cape Cod.

Kimberly Ridley is an essayist, science writer, and author of award-winning nature books for children, including *The Secret Pool* (2013) and *Extreme Survivors: Animals That Time Forgot* (2017). Her articles and essays have appeared in the *Boston Globe*, *Christian Science Monitor*, and *Down East*, and she shares her love of nature with children and adults through her writing and teaching. Forthcoming books include *Wild Design: The Architecture of Nature* (2021) and *The Secret Stream* (2021).

Betsy Sholl grew up on the New Jersey shore and has lived in Portland, Maine, for thirty years. She is the author of nine books of poetry, most recently *House of Sparrows* (2019), and served as the poet laureate of Maine from 2006 to 2011. She teaches poetry in the MFA in Writing program of Vermont College of Fine Arts.

Deb Soule founded Avena Botanicals Herbal Apothecary in 1985, working as the farm's primary herbalist, herbal formulator, and biodynamic gardener. Author of *Healing Herbs for Women* (2011), *How to Move Like a Gardener* (2013), and *The Healing Garden: Herbal Plants for Health and Wellness*

(2021), Soule shares her thirty-five years of growing, gathering, and preparing herbs to inspire others to weave gratitude and respect for plants into all aspects of growing and preparing herbal medicines.

Kim Stafford is the author of a dozen books, most recently a collection of poems, *Singer Come from Afar* (2021). Having taught writing at Lewis & Clark College for forty years before retiring in 2020, he now teaches and travels to raise the human spirit.

Alireza Taghdarreh of Mehrabad, Tehran, Iran, made his first trip outside his country to the United States in July 2015. He spoke at the Thoreau Institute at Walden Woods about his ten-year devotion to producing the first translation of *Walden* into Farsi, followed by Emerson's *Nature*. Taghdarreh talked about the connection between American transcendentalists of the Thoreau-Emerson era and classic and modern Persian poets. He learned English by watching English and American movies and TV, and later, after Iran's cultural revolution, by reading old English newspapers and magazines.

Doug Tallamy is a professor of entomology and wildlife ecology at the University of Delaware, where he has authored 103 research publications and has taught insect-related courses for forty years. His books *Bringing Nature Home: How Native Plants Sustain Wildlife in Our Gardens* (2007), *The Living Landscape* (2014), coauthored with Rick Darke, and *Nature's Best Hope* (2020) were published by Timber Press.

Landscape architect **Thomas L. Woltz** has forged a body of work that integrates the beauty and function of built forms with an understanding of complex ecological, cultural, and engineered systems. As owner and principal of Nelson Byrd Woltz Landscape Architects (NBW), Woltz has infused narratives of the land into the places where people live, work, and play, deepening connections between people and the natural world and inspiring environmental stewardship.

Acknowledgments

Thanks to Jan Hartman, who had the original vision for this book, to Kristen Hewitt at Princeton Architectural Press for her incisive editing, to the contributing writers who enabled me to see the natural world through their eyes, and to Ralph Waldo Emerson whose words continue to inspire us.

Published by
Princeton Architectural Press
202 Warren Street
Hudson, New York 12534
www.papress.com

Printed and bound in China
24 23 22 21 4 3 2 1 First edition

ISBN 978-1-61689-986-8

Editor: Kristen Hewitt
Designer: Paula Baver

Library of Congress Control Number:
2020950585